HEAVEN'S CALL

by Roger Dawson

DORRANCE
PUBLISHING CO
EST. 1920
PITTSBURGH, PENNSYLVANIA 15238

Dorrance Publishing Co
585 Alpha Drive
Pittsburgh, PA 15238
Visit our website at *www.dorrancebookstore.com*

ISBN: 978-1-6480-4643-8
eISBN: 978-1-6480-4661-2

Chapters

Addendums

Chapter 1

Into the Abyss

I felt fear stab me deep as I saw the expression on my wife's face when she took that call near year's end in 2012. On the phone was my daughter, Jamie, calling about something awful. My wife Debra put the phone down and looked at me sadly, saying sit-down, Jamie has something bad to tell you, but I already knew what it was, as she handed me the phone.

It was 15 years earlier to the week, just before Xmas 1997, I had the same horrific call about my wife's 20 year old son, Sean. He had been away for the weekend at his best friend's 21st birthday party. There had been a fire in the middle of the night at the friend's apartment, and two bodies were found. My call then was from the other mother, her son's body was identified, but the other body could only be verified thru dental records, and she tearfully asked me if we would give Sean's Dentist's name to the police?   But others at the party earlier had told her it was Sean, they knew he was staying-over when the others left in the wee hours.

Back then I had to tell my wife the worst thing any parent could ever hear, her child was dead, he was just 20, they were so very close. Tragedy had already slammed her and her two kids a decade earlier, her first husband died from cancer at age 43. I met her a few years later after my first marriage ended. Her family had trauma, now they faced devastation, too much for anyone. Debra would be home in an hour, I rehearsed what words to say over and over, but that didn't help.

When she arrived I asked if she heard from Sean today, she grasped the implication at once, my voice stumbled into saying there was a fire but I put off telling her the need for his dental records, too painful an image. We entered a night of screams and tears, I cry again now just writing about our tragic shared fates, I already knew, as I took the phone from Debra, it was my turn now for the same nightmare. I was numb as she spoke, my son, Zack, had died about dawn in his bed, he just never woke-up.

But it was not unexpected. We had fought my son's addiction on-and-off for many years. Zack's substance abuse had started in college where he played in a band and they got access to OxyContin pills, an easy, preppy, first step into the hell of opiate addiction that has swept the nation ever since. He had been in a cycle of drug use, treatment and recovery for over a decade. In stable periods he returned to the family business- ironically, creating and managing programs for at risk youth. He was deeply involved with helping other kids, a leader, earning much respect. I prayed as a spiritual person, and asked for help for my son, and for all the kids we served. I secretly hoped, even expected, that our work would protect Zack. After all, providence seemed to help us so far with much public support for our work. Our family needed Zack to be well, and I had faith up to then that he would be.

Zack's treatment programs over those years were many, 9 different programs from high school on, he had reached age 30. When health insurance ran out, I used my savings to fund even more treatment, hundreds of thousands spent in total, yet it all had failed. My career had already taught me the high failure rates of drug treatment, though my work was mainly in the wider fields of alternative education and juvenile justice, giving me awareness of most youth programs. I had assumed Zack was just like me, he was so in many ways. In college, my friends and I had dabbled a little into the "hippy culture" popular in America in the late 60's into the 70's. For half my college years we all danced to Motown and drank beer, but by 1969 we all wore bell-bottoms and grew our hair longer (a full 2" below the collar), and listened to Frank Zappa & Mothers of Invention music, what rebels we were then. And some smoked pot, though beer was never abandoned. I had tried a few drugs but never liked feeling "spaced-out." I had a few friends that slid towards addiction but I knew where to draw the line, they were in trouble and I was not like them.

So I had thought Zack would follow my path, stay within bounds. But I was mistaken, maybe also now about a lot of things I believed up to then.

Our whole family had tried hard to help Zack, Jamie and her mom, Paula, even lived with him at my winter home in FL for many months, setting up more treatment there, while I still worked up in PA. They helped him finish college, he did well for awhile with them by his side. My former wife and I have remained friends, maintaining strong relationships with our two kids, who became part of a larger family.

So we all tried hard to save Zack, but now he was dead, my faith in myself died with him.

Chapter 2

Learning How to Help

When I finished college in 1970, my first jobs using my psychology degree were all good learning experiences. The first lasted 3 years as a middle school teacher at a residential school for emotional disturbed youth. So many of these kids had little family contact and they craved relationships, someone to care about them. I had empathy for them having faced much stress myself as a child. I grew-up with a mentally handicapped mother who only attended elementary school, I learned little from her, though she was pretty and happy in a child-like way. My dad was kind, shy and withdrawn, that I could accept, though his disinterest in me I could not.

By my teen years, he lived at times away with his girlfriend and that made mom even more anxious and disturbed, with strange, compulsive behaviors. I didn't like having friends visit me at home and seeing this for themselves. I felt shame, especially when I got old enough to understand how other adults reacted to my mother, and I stuffed that pain deep inside, it has never left.

Besides my parents, my family then included my genius older brother with 150+ IQ, and photographic memory, I'm told he could read a book and remember it all. But in my opinion, he was also on the mild end of the autism spectrum, that meant he almost never spoke to, or played with me, that didn't bother me. By the time my brother left for college as I entered high school, dad was often not living at home either, and mom got much worse. She was left only as my problem, but I was not a good son, I could offer her little, I was too self-centered, maybe that was necessary, and just held on until I could leave too.

In my teaching job at the residential school I got to know many kids who had it much worse than I ever had, so my childhood wasn't that bad after all compared to their lives. Mostly, the rest of my family had just left me alone and I raised myself, some positives came from that like being independent, as well negatives, like distrust of some in authority positions.

Working with those kids at the residential school, I learned perhaps for the first time, to care about others, I got attached to some kids, bonding, with empathy for their feeling of abandonment, I knew that despair too. My caring for others slowly became a trait, fed by practice, I especially watched over those weaker kids picked-on by bullies. I had been through that myself years earlier, though in a different way than most kids, having been abused at times by those in the community who belittled my family, that happened too often. In my first job I was far from an ideal worker, yet felt good knowing for sure that I helped some kids. Over time, fate was to create more opportunity for me to try to do that for many others.

Drawing on my love of the outdoors, I often took these needy kids hiking, my love of that had been nourished by my wonderful earlier experiences in Boy Scouts from ages 11 to 15. Troop 203 was a large Boy Scout Troop with over 100 kids, even a marching band in which I became a lead performer, marching proudly in many parades, I loved to show-off, perhaps it healed my inferiority and anxiety. Every Memorial Day, our band marched in the parade that ended at the Veteran's Cemetery on Olney Ave, Philadelphia, where I would play taps by myself in front of thousands. More positives from Scouts were the many surrogate fathers, with some supportive role models. I really needed that then, didn't we all?

And I especially loved those thrilling camping trips that were to become a continued part of my life, my career, and shared with my future family. I discovered nature heals, but had no idea back then our family would someday be able to offer that healing to so many.

My second job for another 3 years was as a social worker for city of Philadelphia public housing. There I learned much about the good and bad of government programs. More importantly, I learned casework, the central helping process of most social services. We learned how to respond to people in crisis, that trust and sincerity mattered, and we leaned about gangs.

We started by playing basketball with the local project youth, the project's indoor BB Court had been locked-up for years, so we reopened it and played ball after school and got to know the kids. We even got them organized into a team and entered tournaments, though if they lost, they wanted to win a fight after, that we stopped. We were close to some, and they respected us as they knew we had actually helped families.

We operated a Food Bank supported by charities with donated foods gathered at schools that we delivered to those in need. Regularly, large trucks delivered tons of Tasty Cake bakery products to our project office that we gave out through the food bank. But I must finally confess here and now about those empty wrappers in my desk trashcan. They were not just from my making sure the treats were fresh enough to give away. We were not like some other staff there who did little work.

We took the project youth camping on my favorite hike on the Appalachian Trail, the Pinnacle, it had a 1,000 foot high cliff and 30 miles long views in 3 directions. Most of them had never been in such beautiful place. I remembered what Scouting had done for me, gladly giving them a little taste, they savored it, I always shared my love of the outdoors, and what I gave to others seemed to be returned in unexpected ways, becoming another life pattern for my future family.

The Pinnacle Overlook on the Appalachian Trail

Chapter 3

A Successful Failure

By the mid 70's, social services across America were in the midst of major changes. Large state-run institutions from past decades, across a variety of fields, were now deemed as both expensive and harmful failures. By my 3rd and final job in 1976, a national movement was well underway to close large public institutions, with the new public policy goal everywhere of "**deinstitutionalization.**" That meant helping people in their neighborhoods rather than concentrating them into large institutions that often made them worse. I had seen institutional harm myself on a small scale at my first teachers job for emotionally disturbed kids. About 100 students lived at that residential school, and daily, especially at night, there were many more unsupervised interactions among troubled youth than with any therapeutic staff, and some were very nasty and harmful, as were some staff there too. The good that came from a few hours of therapy, for some, was overwhelmed by the more frequent harm from negative interactions with others.

The new model to replace institutional care was called **community-based services**, a new national strategy in the juvenile justice and mental health fields. Few really knew what should come next with these nation-wide reforms, we had to figure that out in the late 70's. There were new youth programs everywhere and I was hired by one, Youth Services Agency (YSA). The judges who supported YSA were powerful, the statewide leaders in both mental health and juvenile justice reforms in PA. As large institutions closed, a state mental hospital called Pennhurst, along with others, and state reform schools like the Bensalem Youth Center, there was reshuffled funding and grants to demonstrate new community programs. I was in the right place at the right time, excited to be part of this, what a creative time for us all in the field of youth work.

YSA, my new employer, got referrals from police, Juv. Court, and public schools to help youth with problems on a 24/7 basis. Our mission was **diversion**, which meant to keep youth away from juvenile court and criminality by placing them into whatever services we could find that might help them in their communities. Especially, we were to figure-out how to help one group called "status offenders" whose crimes had now been decriminalized following a landmark Supreme Court ruling (in addendum A). Those offenses were running away from home, incorrigibility, and truancy, all relating to the offender's non-adult status. These kids could no longer be arrested, instead, in our region, they were sent to YSA and other new youth service bureaus across the nation.

YSA served 1,000 youth a year back then, with more each year through the 1980's. We used 100 different community programs that we thought could help kids. Big Brothers and Sisters had specialized court-funded programs for just our clients. As well, other new in-home therapies, day treatment programs, high-risk youth job training, family counseling, and alternative education programs, they all prospered if effective.

Resistant youth were my personal priority, if the referrals we made had helped clients, wonderful, we had done our job. But if not, rather than giving-up on kids who kept re-offending, I felt free to try new ideas. Those extra efforts fed YSA's growth and my career, as we created new community-based services. We had gained political support, had powerful local supporters, but I was also self-destructive, being independent, I fought with those authority figures I didn't respect, I made enemies easily, I was good at that too. I coined the phrase, "friends come and go, but enemies are for life, " some truth in that.

By 1984, YSA changed from a small county agency to a mid-sized private-non-profit to support our adding new programs to fill gaps in the new community-based system. From then on, YSA's size and funding steadily grew by 15% a year for 15 straight years. From a $250,000 budget in 1984 and a handful of staff, to the annual budget of $24 million by 2005, reaching 500 employees over 20 sites, and funding from 3 states. Creativity, low costs and good outcomes were the reason.

We increased outdoor adventures by going backpacking most weekends year-round on the Appalachian Trail (AT). We opened two campgrounds with indoor facilities about 25 miles apart to support overnight backpacking, and the camps grew into 7 days a week residential programs for up to 175 youth. Eventually we built and operated ropes courses with rock climbing walls adding to our ability to provide Adventure Challenge Therapy (ACT) to most programs. These stressful activities produce positive emotions in reaction to the fear inherent in these challenges, to be explained in detail later. Over time, these activities promote group bonding, my goal was for YSA staff to become positive role models, lead by example, thus teaching values, very effective therapy.

We provided adventure therapy, eventually, to tens of thousands of youth using our 20 locations by 2005. Over my career, about 35,000 at-risk youth went backpacking on the AT, with about 150,000 total youth served from 1976 until 2017, when I fully retired. Add another 100,000 if we include Barn Nature Center visitors. More details of YSA programs presented in Addendum.

My own kids from junior high on were very involved in these adventure activities, often participating, it was a part of all our lives. Over the years, Zack and Jamie became YSA leaders, certified by Outward Bound and others, with certifications in rock climbing, kayaking and first aid, etc. They even helped on our 78' square-rigged tall ship with real cannons. Goggle tall ship "Lioness" to see it on U-tube at:

https://www.youtube.com/watch?v=XtiI6CXPMCA

A Lioness picture is in addendum. We participated in mock tall ship battles with other vessels from Baltimore to Phila., some televised. Youth doing well in our programs were rewarded with weekends on the ship, I knew having more rewards for good behavior than punishments for bad is essential but rare in programs. We had done well, felt successful, and had pride from these positive outcomes.

Years later, YSA even got the attention of the Whitehouse, in 1999, President Bill Clinton was considering a site visit to our impressive AmeriCorps Barn Project for a dedication ceremony, though I screwed that up quickly, a story for another time, it never happened. YSA prospered, reaching its peak from 2003 to 2008. But for both YSA and my son, success was not to last, after a multi-year stable period, Zack started to go downhill again around 2009. By early 2012, my son had yet again completed months of seemly successful inpatient treatment at a top ranked program. As aftercare, they sent him from PA to a halfway house in Minnesota.

As always when sober, Zack impressed everyone, a natural leader with extensive training in social services from his work at YSA. But he was near the end now, I knew it, we all did. Thankfully he enjoyed a beautiful sunset to his life in his new job managing a group home for severely handicapped adults in MN. He would play his guitar and sing with the blind and handicapped adults he lived with at the group home, he was loved there too. Yet after 9 months, he came back home to PA for his cousin's wedding in fall of 2012, and he seemed not right to me. As he left to return to the mid-west, I thought while hugging him as he got in the car to the airport, I may never see him again, in two months, sadly, that came true.

I sat in the dark that night he died, crying and thinking about it all, my life's work, how I failed with my son, now gone forever, devastated. I accepted the obvious conclusion, I was just a successful failure. Could I help anyone, if not even my own son?

My imagined conversation with God still haunts me. Years earlier, as we first struggled with his addiction, I prayed for him with awareness that our new adventure-based therapy might help him to recover. He was physically and emotionally very strong from rock climbing and overcoming the inherent fear in that, it satisfied his euphoric cravings.

In simple terms, one can get high in a healthy way thru adventure without using drugs, thereby satisfying the cravings rooted in the brain of both emotions and addiction. Rock climbing was among Zack's favorite pass-times and gave him an emotional high. Stressful activities directly affect brain neurotransmitters, the chemicals that cause emotions, the same that are stimulated by illicit drugs. These neurotransmitters try to maintain equilibrium, to stay in balance, that process is called Opponent Process Theory as proposed by Richard Solomon of University of PA. I had published research in a national journal suggesting it might help treatment resistant, high-risk youth. I had the thought then, and always with me since, that **God** had said to me:

"OK Roger, sounds like a good idea, but lets test it first on your own son." But that test for Zack had failed, not what I expected, so I lost faith in my work, my ideas, feeling spiritually abandoned.

But providence wasn't done with our family just yet. It took only a day more for that to be revealed in a startling way, a special gift.

Barn Adventure Center about 2001,
one of our students leading a group through the low adventure course

Chapter 4

The Call Down a Rabbit Hole

Zack had died early Sun morning in Saint Paul, MN, but it took until evening for word to reach his mother and then my daughter. Jamie then called me, and later that night some close friends too. In a few days Jamie called me again, excited, and with a bazaar, unimaginable story of events that had just occurred. As she talked with her friend Laura over her cell phone, Jamie heard a third voice interrupting, loud at first, strange, not understandable, then becoming very familiar, it was Zack's voice, he said: "I'm better here." At first she suspected it was a cruel joke, but how could someone interrupt a cell phone call, and not many yet knew he had died over 1,000 miles away.

Laura had called again and Jamie didn't answer, so the message went to voicemail. Laura offered to come over and spend the day with her, but at points in the voicemail recording, as Laura's voice pauses, perhaps to breathe, Zack's voice replaces hers. He says about 5 sentences, low and raspy, but it's clearly his voice. You can hear the recordings for yourself at the Dropbox links at chapter's end.

Not all words can be understood, just parts of the first two messages. He says he is hanging out with Sean, (this is 15 years after Sean died). "Its so crazy" … (presumed he says I'd show you, but inaudible) then its clear he says "but I don't know how" and "will visit you at the house," He says he is better, and he ends with "its sooo perfect".

Amazed, confused, stunned, frantic, grateful, we felt all that and more, beyond the capture of words. He had contacted us just after he died, and we had our proof in voicemails. We got 7 messages in 10 days, but only a few are somewhat clear, the rest the same voice but words all garbled. All are his voice, 4-5 sentences each. He says in the 2nd message "seeing many people, they don't know much … sorry to have left the family, love you, I'm better".

Frankly, over the years since, we could care less if anyone believed us or not, our family has zero interest in trying to convince anyone. We just feel at peace ourselves knowing there is an afterlife and our loved ones still exist, what a gift. Their consciousness continues as pure energy beings, maybe as part of universal energy fields nearby. But they do visit and seem to know what's going on here, somehow they can find us. A few years later we would also get a phone camera photo of either Sean or Zack as a semi-transparent boy with something very special about it soon to be discussed.

We just want you to know what happened, we feel a duty to tell this story so this miracle is shared and not lost, that's important. We cannot conceive of how it could have been faked, we asked others to check that. There was a small group of friends and relatives who witnessed some of these events, little presented has a single person source. Everything as accurate as possible with no embellishments, these events really happened just as described here.

We contacted several ESP groups and sent them the recordings to check and they found no fraud. Unexpectedly, events soon connected us to some top U.S. ESP experts, formerly with the military. The Army, CIA and NSA had funded extensive research of the paranormal from 1970 thru 1992. I met with some who worked in those secret projects and are now free to discuss their stories, and the CIA released their research starting in 2003. Some military related ESP experts afterwards became part of a psychic research institute in VA discussed next. The military did this research because the Russians were using psychic spying on us, and the CIA needed to know if it worked.

The last message from Zack was just before his funeral. I arrived an hour early, and as I drove up Jamie was there already and excited. She just got another phone message saying "be careful what you say." That was relevant as just the night before she and I discussed that we expected an addict friend of Zack's might attend his funeral. We blamed him partly for Zack's relapse. Jamie and I agreed if he came we should ask him to leave. Zack must have heard us, and his last message was his answer, he didn't want us to do that.

About 400 people came to his funeral, he was surrounded by photos of his life, his artwork, and by so many people who cared about him. I wanted to tell them all about this amazing event, and that Zack's consciousness still exists, proof of life after death, but I didn't, and I cried thru much of the service. Over time, as we try to tell others, we find it is not easy, its a long complicated series of events. Some we told thought we were so in grief we must have imagined this, others that we were just plain deceiving them. While some have no trouble accepting it fully. The details and proofs all need to be absorbed as a whole to fully comprehend these events.

Jamie and I got a special gift as we entered the funeral home that day. There waiting outside the door was a young man in his mid-20s whom I didn't recognize, and he was crying. He told us without our asking, how much he loved Zack. Inside Jamie told me the story, he was a student in our Barn alternative school years earlier, and was kind of kid that bullies picked-on. We had our share of aggressive students, though we also got those who needed some intense help for other issues, as with this student. Zack had given him much support and esteem, teaching him to be leader to manage the Barn's climbing walls and ropes courses for both other students and the public. We always sought to develop students into leaders to help us provide adventure to the public, giving them leadership and positive roles in their community, that is effective therapy too. This young man was proof of Zack's impactful life. He helped us bear the crushing grief we all felt at his funeral, and our grief remains nearby always. Perhaps his contact following death was a gift to our family to help us continue our work by assuring us that Zack and Sean still exist.

After the funeral, Jamie, Debra and I decided to get some training in meditation; maybe we could find a way to have more contact with them. Proof of an afterlife is important, that's a goal with this book, and writing it became even more important after Zack contacted me directly about doing that just 7 months later, he wanted me to tell this story, this among the many strange events to occur over coming years to our family.

Zack's voicemail recordings are accessible to readers online at Dropbox, here are the links:
https://www.dropbox.com/s/7hhj566n98l992y/Zack%20Video%201.mov?dl=0
https://www.dropbox.com/s/8eo3kqsf24gd2sf/zack%20video%202.mov?dl=0
https://www.dropbox.com/s/ihnive1evppacig/zack%20video%203%20.mov?dl=0

Chapter 5

"Signs" Lead to Monroe Institute

Dr. Eben Alexander is a neurologist who became very ill and went into an extended coma and nearly died. He had an intense out of body experience during the coma and wrote best-selling book about it called *Proof of Heaven*, published just before Zack passed. He had much credibility given his credentials. Jamie went to one of his public speaking events a few months after Zack's messages. Many attended but at the end he said he only had time for 2 questions, many raised their hands, of course he chose Jamie. She asked what helped him the most to understand his experience, he said he attended the Monroe Institute, it gave him insight and similar experiences.

We looked Monroe up online, and immediately another coincidence occurred. My wife's friend is a travel agent who arranges trips to spiritual and other unique interesting sites. She told Debra in February she had already booked a group training at Monroe Institute for late May, and Monroe had just 3 spots left in this training. Hum, to us these two events might be more than a coincidence, so we registered for the Monroe Institute "Gateway" training in meditation and ESP in the Blue Ridge Mountains of VA. Of interest as paranormal events were increasing for our whole family, even grandkids. More were about to occur to us during and after the late May 2013 training.

The agenda at Monroe is group lectures and discussions for about 45 minutes, followed by a return to our bedrooms for individual 45 min. meditation practice sessions. (And Yes, I do now admit to an occasional mid-day nap instead of practice), I was never good at meditation. At the next group discussion people shared their meditation experiences if they wanted, this was repeated 2-3 times per day with longer evening presentations. They were all interesting and it's a serene and beautiful campus. The trainers have much expertise and the audience was composed of interesting people too.

Each trainee's bed was setup as a semi-isolation chamber with a few beds per room. Curtains gave darkness and privacy, and stereo headphones were part of each bed. The audio via headphones is part of meditation instruction, with sounds that also affect your brain wave frequencies. Beta wave frequency measured in Hertz (HZ is cycles per second) is from 12 Hz to 30 Hz, for normal consciousness, then next is Alpha brain waves, a relaxed state of around 12 Hz, followed by Theta, the sleep state of around 7 Hz, and lower still is Delta, in which we may have paranormal experiences. Monroe developed Hemi-Snych technology to synchronize brain waves into deeper levels of consciousness that can improve meditation.

Theta and Delta brainwave states are the goal of meditation, below 7 Hz, and it can take much practice to reach the lowest state of 4-6 Hertz (Hz) or less. We were told even Buddhist monks take many years to easily reach this lowest level of brain wave frequency. Monroe Hemi-Snych using stereo headphones, broadcasts 104 HZ in one ear and 100 HZ in the other. In the brain the 100 HZ wave-lengths from each ear cancel out thru wave interference, leaving just the desired 4 HZ, the deep meditation state believed to help stimulate visions and ESP.

Before attending Monroe I was skeptical and assumed it would be like other "New Age" retreats I've known, that too often for me had little science supporting their teachings, as well with some naive devotees. But that's not the case at all at Monroe, some of the trainers, as said before, were former U.S. military research participants. Monroe had once been a state-of-the-art research lab with sensory isolation chambers with electronic sensors and data recording equipment. Google Tom Campbell, a physicist formerly with NASA, he was the Institute's lead researcher in its research years, he has easy to understand U-tube lectures available. Monroe was a pleasant surprise, not what I had expected. They had proven the existence of telepathy, as have other scientists since, and the CIA evaluation of Monroe's Gateway training is discussed in the last chapter. The military research connections with Monroe are beyond just people who later came there after their service, the CIA evaluation of Monroe hints at deeper connections.

Scott Atwater had been the Institute's CEO after retiring from the military research projects at Fort Meade, MD, and Joe McMoneagle, also on U-tube, was one of our lecturers. He and Scott were part of project Stargate (formerly called project Grill Flame and other names). Most of these programs have now been declassified. Remote Viewing research sponsored by the military was also evaluated by the prestigious Stanford Research Inst. (SRI). I had seen a TV show a while ago after Monroe showing Joe doing a remote viewing test at SRI in CA.

On a lighter note, these US Army research projects were "spoofed" in a funny movie with George Clooney, Jeff Bridges, Ewan McGregor called *The Men Who Stare at Goats.* Though a comedy movie and very funny, it is much different from the reality on which it was based. The US Military was most concerned that the Russians were spying using one type of ESP called remote viewing. Some meditate, then using their mind only, view far away places around the world or beyond, obviously useful if true.

Bob Monroe, the founder, was a high level broadcasting executive before starting the Institute, he had life long experiences with out of body ESP events. I had a few of these myself, from age 8 into my teens. Maybe 15 or more times, which I assumed then were just lucid dreams, until I learned more at Monroe's training. These start with my conscious awareness that I was floating over myself in bed, I could see my body below, or sometimes I was floating outside my bedroom window. I could not see or feel my body, and my vision was fishbowl like, clearest in front of me, less so towards the edges. I could go up or down, left or right by thinking alone, exciting, I loved it, and was not afraid. But would never stray too far, just floated around the yard maybe 10-15 feet in the air, again, upon waking, thinking these were just repetitive, pleasant dreams.

At Monroe Jamie had an especially significant experience. She was leaving a month after her Monroe training to go far away to Borneo to study primates for awhile. After training on how to meditate, we were given an exercise to seek answers to a question, we were excited, exactly what we had come for. I had no success at this but Jamie did, after all, Zack had contacted her first and foremost, they were very close

In the group discussion that followed that exercise, Jamie shared that she had seen Zack but he didn't speak, just smiled at her while pointing at her seemingly pregnant stomach in her vision. She felt this was because he knew up to then she didn't want to have kids herself, and her interpretation was that Zack now wanted to encourage her to have a baby. However, a few weeks later, after we were back home, she had a very lucid dream in which Zack spoke: "YOU ARE PREGNANT" he informed her. She got tested the next day and it was true, and she had been barely weeks pregnant when at Monroe when Zack first pointed to her stomach. Now she had learned she was pregnant days before leaving for Borneo thanks to Zack. She had "morning sickness" the whole time away.

Almost 9 months after visiting Monroe, my beautiful grand daughter Kia was born in Feb. 2014. Kia, today age 6, as well as our two other grand-kids, Marly and Prescott, have all helped fill our family's emptiness from the loss of Zack and Sean, as does their spirit visits, they hang-out with the grandkids too as shown in an enclosed photo.

Though I never could improve my meditation abilities, I did have one ESP success while at Monroe. In a in remote viewing test the instructor put both latitude and longitude numbers on the blackboard marking a precise spot on earth, and asked the group to each draw what they saw there. The numbers meant nothing to me or others, I saw in my mind just shapes, not a place, I drew an arch shape next to water, I even wrote water by the squiggly, wavy lines I drew next to the arch.

Of course, we were later told the target was the St. Louis Arch monument. Interestingly, the next training the week after ours was in remote viewing, but the Monroe staff had earlier said the training was already full when someone asked. Now the instructor said they had an opening. Hum, I thought, maybe that was an offer for me, and maybe some others, after he viewed our correct answers. I wasn't able to stay longer, and my conclusion was that Zack had given me the answer, as he was there with us at Monroe.

Monroe had been eventful for us, besides Jamie's experience, my wife felt contact with Sean, and I also learned about my earlier out of body experiences. The training helped us understand that there was good research proving ESP. We were told their belief that all people have these skills, there are more than 5 senses, but intelligence and suffering are two triggers that increase ESP, and my family had those triggers.

Good to know as the whole family entered a time with more ESP events. Many, but not all, of our them are positive, and I won't cover them all in this book. My most recent, in Feb. 2020, was very funny, a voice from another room asking me a question, more about that soon. Note, a few family members don't share all events, and a few names changed for privacy.

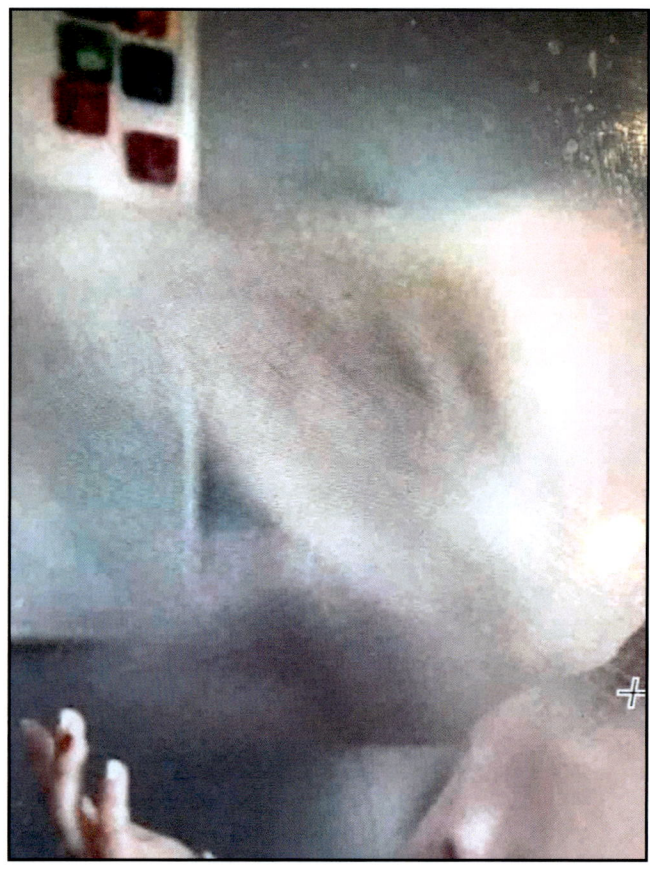

A Visit by Sean or Zack?

Chapter 6

ESP x Three

All 3 generations of our family have extra sensory perception (ESP) events. Debra and I have been together 27 years, each with a son and daughter from our previous marriages. Besides the loss of our boys, our two daughters each had kids. Alyson, has Marly, age 12 in 2020, and Prescott age 16, while Jamie has Kia age 6 today. For some in our family their ESP events began in childhood, for others, mostly since the paranormal events started in 2012.

For myself, they have always been a part of my life. I have thought about my first memory, as it is, perhaps, a memory of my birth, though doctors might say that's impossible. This earliest memory has been with me since I became self-aware, I even remember lying in my crib and thinking about it, I was old enough to climb in and out of the crib myself, maybe age 2-1/2 or 3. I remember it was easy for me to slide the crib side down and climb out of it, but once back in, I could not raise and lock it again. In my memory, I opened my eyes to a blurry world of light, I could not focus my eyes to see anything as if I didn't I know how, then I saw something white coming close to me and I recall squinting my eyes and it suddenly came into focus for just a second. I got a quick glimpse before returning again to all blurry vision, returning quickly to unconscious asleep.

What I saw was the face of a middle-aged man in a white lab coat leaning close to my face with one of those round shinny metal reflectors they used back then to focus sight. It covered his one eye.

From age 6 on, I learned to sleep with the covers mostly over my head to get away from seeing things in my room. Often just twinkling lights, not scary, but sometimes shadows, faces and sensations of being touched, never inappropriately, but feeling hands on me, that was very scary. So I learned to hide under the blankets, using the primal denial trick that if I don't see it, then it's not there.

I discussed already the recurring dreams that happened thru my childhood, with awareness of floating and looking down at myself. Those out-of-body dreams stopped about age 18, with just one more very dramatic one that happened recently. But at life's end, I think we all have one last out-of-body event.

I conclude minds are not just physical matter, but, as we will soon review, science tells us matter is an illusion, beneath apparent particles, such as atoms, reality is composed of energy waves in constant motion. As Nickolas Tesla said, "if you want to understand the universe, think of vibrations and frequency", those are characteristics of waves. I've read that Tesla and Einstein both attributed some of their insights to ESP in their dreams. Matter and energy are 2 states of the same thing. Out of body consciousness may be similar to those traveling balls of energy called orbs. A photos of orbs above my wife and at the Barn are in addendum, along with other pictures of family ESP events that will soon be discussed.

My most frightful paranormal event took place at age 20 at a summer job while in college. I worked at a large 5 star restaurant and resort on Rt. 30 in Downingtown PA, along the old main road from Philadelphia to the state capital. The resort owned an old, large, colonial style house in fair condition used for employee housing and located a few miles from the resort. It was once called the Kings Arms Inn, and had a wall plaque saying circa 1698, it is listed as the oldest standing building in Chester Co, PA. Being a good worker, they rewarded me with a cheap room there for the summer. The restaurant had over 50 staff, many college kids like me, so I had a fun summer.

In my room on a bright sunny day, I leaned back in bed and glanced at my clock, it was 1 pm. I closed my eyes and immediately was swept with an intense fear, with no idea why, I must have sensed what was about to happen. I opened my eyes and out of the wall came a black shadow figure with clearly seen head and arms out-stretched towards me, with torso not quit to the floor. As it glided towards me its outline was wavy, not fixed. I put my arms up over my face and closed my eyes long enough to think it won't be there when I look again. A second later I screamed as it was still there and coming closer, then slowly dissolved about 5 feet away. The clock showed not even a minute had passed.

I have learned since that shadow figures are somewhat common, as are EVPs (electronic voice phenomenon, which describes Zack's phone messages). Take a tape recorder to most any large graveyard and ask questions, eventually you may get answers, these recordings are called EVPs.

Another paranormal event relates to Debra being a member of a large international online prayer group and she had two good health outcomes associated with this group, coincidence or not, I give them credit. Debra submitted the names and health issues of the ill and thousands in the group prayed for them. The first was a co-worker who in his early 50's developed lung cancer. Tom was a gruff ex-marine who managed some of our adventure and community service programs thru YSA's PennServe and AmeriCorps grants, I worked closely with him. With a drill Sargent's demeanor, he was effective at keeping our youth and staff productive and getting large projects completed. Especially our restorations of two publically owned historic Barns, one from 1732 and part of the county's poor house, the other from the 1850's. Tom showed us an X-Ray of his lungs, one was completely clouded by a large tumor, over a month later and after the prayers, his next X-Ray showed it was totally gone, we believe we had witnessed a miracle.

The other involved my daughter Jamie, and a Mayan child in Belize where Jamie did a field placement as part of her graduate school program in which she spent months studying wildlife. A 6 year old boy had an acute liver infection and with no further treatment available there, his prognosis was very grim. Jamie asked for help and I agreed to fund his plane trip to Miami for emergency medical treatment. Somewhere during the boy's journey north, he too was cured after placed on the same prayer list. However, we where not along with him, just indirectly told of the cure by others, so not certain of all facts.

Jamie had spent much time traveling as a volunteer in Central and South America in between stints working at YSA. Every few years she would go away for an extended 3-6 month trip, traveling to volunteer in animal care programs, and at times Zack went with her, our family always needed these getaways.

While in Peru, Jamie again helped rural indigenous people, she had noticed that some children and adults living high in the mountains had black marks on their feet. She was told the black marks were due to frostbite from them not having warm socks as the winters high up in the Andes Mountains are very cold. Jamie then maxed-out her credit card buying car-loads of used warm clothes she found in the city of Lima. Filling several cars, she then went back-up into the mountains and gave them all away to kids who were in need. Jamie has earned spiritual grace, maybe her kindness repaid by Zack's contact?

My wife has had many ESP events too, mostly she hears voices, generally in that semi-conscious twilight state just before actual sleep, or when she awakens in the night due to bedroom noises, which happens too often. Sometimes I awaken with her and hear the same voices as she does, we can hear several voices seeming to talk among themselves, but we can't quit catch the words. Sometimes she clearly understands what's said, "we love or miss you", sometimes the opposite, "we don't like you," she thinks these are from people in her past her lives.

After our Hemi-Snych training, spirit voices occur more often, even during the day, we sometimes hear a voice just calling our names from the next room. I was home alone and just got into the shower and closed the curtain when a women's voice in the hallway called "Roger", I laughed and called out: "Thanks for that … right out of the movie Psycho." No response, guess no sense of humor. The most recent, and funny, while alone, upon waking-up and about to get out of bed, a whole sentence from a polite female spirit in the living room asked "Roger are you dressed"? I laughed and answered her no, she didn't check.

A full body apparition of Debra's mother has been seen too. Debra and her sisters had a yard sale while getting the house ready to sell after both her parents died a few months apart. The upstairs bedrooms were full of her mom's stuff piled on beds, and her mom loved the color purple. Sitting outside the front door, Debra and sisters saw an upset shopper coming from the house, she said who was that older woman upstairs who didn't want me in the bedroom, Debra asked for a description, the answer "she wore a purple dress". Of course there was no one in the house.

ESP events of our grandkids are interesting, but sometimes worrisome too. Prescott is a talented baseball player with many awards, he is on the all-star statewide traveling team that entered a national tournament. Though not among top ranked teams there, they were seeded to play against the # 6 ranked team in the nation, who even taunted them before the game. Prescott's team beat them 11 to 0 with Prescott the MVP.

A few years ago in his room in the morning he heard a voice, he assumed Sean's, say to him "you have baseball tryouts later today" which he didn't know about. He ran downstairs and asked mom who responded the tryouts for the all-star team are tomorrow not today, but then she checked, and actually it was that afternoon. The spirit voice prevented Prescott from missing tryouts. Maybe "Angels in the Outfield", or in his case the infield, as he plays 2nd base. Spoiler alert, we may have an Angel sighting soon discussed with picture.

All 3 of our grandkids have ESP events. Kia, now age 6, has seen shadow people on multiple occasions at home, Marly, years ago about age 4, was driving with parents past a graveyard, she asked "who are all those people standing over there". She described them as dressed in "old time" black and white clothes. Of course, the parents saw no one. Marly also hears voices as does her grand-mom. A spirit asked her a question, "are you happy"? Marly asked who are you, she said I'm your unborn sister, mom had a miscarriage before Marly.

But the ESP events with photos are the best, some included at Book's end. The first picture is about 5 years ago at Alyson's house, where Marly had a sleep over party. The photo is Marly's friend who had her face playfully smeared with white makeup. Alyson took a picture with her phone camera, and saw something in behind the girl as seen in the photo. Note, I blacked-out the girl's face for privacy as well as faces of some kids in other photos.

Behind the girl and between her and the wall, and turned partially facing the wall, is a boy's partial torso. He is semi-transparent, and blocks out some of the grease-board on the wall, but not all of it, and unlike a shadow, he has color and casts a shadow on the wall too. You can see a reddish colored shirt, a skin color face profile with nose and eyes, and brown hair. What I have seen by blowing it up is a hat that looks like a baseball cap, Sean often wore one, and of course, we already think Sean follows Prescott's team.

Most interesting is the white area on the boy's back facing towards the girl as she looks at the camera. When blown-up, it looks somewhat like feathers with quills, as well, the outside white edge is more blurred as if moving, while the inside white area is clearer, could the white "thing" on his back be moving, an Angel? Draw your own conclusions. Photo is a blowup of the spirit boy, too bad this was an older phone camera.

But there is even more with this ESP event. Alyson emailed the picture to me soon after taking it, and I sent it on to Jamie who lives 150 miles from Alyson. By coincidence, Jamie and her mom, Paula, had an appointment with a talented psychic just a day or two after this photo was taken, it had actually taken Jamie and her mom two years to get this session with Elizabeth Herrington. As Elizabeth started the session, she said she knew that Jamie had just received a very strange picture, and Zack was involved with that. However we don't think it is Zack in the picture, as it looks very much like Sean at age 7-8.

Here is my theory, pure conjecture, of why Sean and Zack are maybe Angels now. When we pass over, as explained by a Monroe trainer, we are met by family to soothe us and ease the shock of transition. My theory is some children have no family, or even familiar greeters, so Zack and Sean have that social work job on the other side. They did that work in life already, so experienced with helping others. So did they earn Angel Wings in Heaven for helping others, maybe also visits back to us? Zack clearly had help to reach Jamie so quickly after he crossed over.

Next are two photos of Marly from a gymnastic meet in 2019. She has been on a team for 4 years and very successful, she now wins metals at most events she enters and just was invited to work-out with the US Olympic team at an upcoming event. These two photos were taken by her mom just a second apart at the awards ceremony after a competition. The top photo is normal but then a split second later the next photo she has no arms. Something is in front of her though she had not lowered her arms or changed her pose, just her arms are transparent in the second photo. Actually you can barely see the outline of her arms still above her head. So strange, in a close-up, there are, seemingly, shapes or faces in front of her, very strange.

The next picture is Debra doing a yoga pose with a perfectly centered, large green orb above her. When magnified the orb has a very detailed structure within, maybe energy lines? Of course you can't rule out more mundane explanations, I try to consider all possibilities. Many colored orbs are shown in the Barn ghost investigation photo in addendum..

Another strange ESP event is my recent out of body experience three years ago, the first since my teen years. It also relates to the issue of past lives, which I didn't believe in until this happened, as I have no such memories, unlike my wife. One thing I was told at the Monroe training in 2013 was to state your intentions clearly of what you seek in meditations. My conclusion about that is to have telepathic or any communication with spirits, they need to hear your words, either stated in your mind, or by speaking out loud. As I've said, I believe spirits are energy beings with consciousness.

A week before my out-of-body event, while meditating, I had asked Zack to help me know if I had a past life, as the Monroe training advised. He heard me by what happened a week later at exactly 3 a.m., a time, by the way, when a lot of ESP seems to happen. While asleep or in that half-awake state, I had a vision, I was floating above and seeing myself in bed, and was instantly aware it was an out of body event, so self-aware during the vision. My wife was asleep next to me, and I heard her voice clearly say: "I had an affair with Major John Wickham of Australia." I jolted upright in bed now fully awake, certain I had never heard that name before, I swear. What just happened, had Debra just confessed something?

Who is this guy Wickham, got up and out to my computer checking that name, the computer clock said a few minutes after 3 a.m. So I searched Wickham and found lots of information on him and his descendants, there are living relatives today. One was a VA governor in the late 1800s, and the only Wickham who lived in Australia was John Clements Wickham, an officer in the Royal Navy, and I then noticed a few coincidences.

Wickham was born in 1798 and died in 1864. He became First Officer of the HMS Beagle and was the 2nd in command during the famous voyage with Charles Darwin to South America during which Darwin gathered evidence used later to derive his theory of Evolution. Wickham and Darwin remained life long friends. Wickham, on a later voyage, charted the coast of Peru, now as the Beagle's captain. He served for years on that ship back in Australia where he mapped coastlines and harbors, naming one harbor Port Darwin after his friend. Darwin gave Wickham a gift, a giant tortoise named Harriet that lived 175 more years, Harriet just died recently in 2006, that is another coincidence linked with our family as you will soon see.

After retiring from the Royal Navy in 1843, he was appointed as both court magistrate (chief judge) and governor of Australia's newly formed second colony, Queensland, with headquarters in Brisbane, AU. He served there until 1857 when he fell ill and went back to Europe, dying 7 years later.

So what did this mean, did my wife cheat on me over a hundred fifty years ago? Was I him back then in a past life? What's the connection to Darwin? Also the coincidences, I have always loved science, from childhood I was artistic and often drew tall ships, fascinated by them, ok that's not much, but also:

1. Jamie lived in Peru for about 6 months in 2006, once back at YSA she began a new YSA program, the first of our two Nature Centers (NC) that were mini zoos, teaching animal care and science education, Darwin would like that.

2. Jamie got a large tortoise for the new NC near same time Wickham's tortoise gift from Darwin died.

3. The week before 9/11, I went to Brisbane AU, Wickham's hometown, to buy a sailboat. I made a deal to import these sailboats in pieces with the plan to have our MD students assemble and finish the boats in Maryland as part of YSA's vocational training for school drop-outs called "Bridges". Jamie managed that program. These multihull sailboats made by the famous John Brown.

4. YSA owned a 78' square rigged tall ship, (see the picture of the Lioness), it looks a lot like the HMS Beagle, and of the same time period. We had many adventures with YSA kids over the years with this ship. I even organized a mutiny, (we didn't kill the captain just stranded her in Buffalo, NY) when we first moved the ship down from Canada in 1997 and she had trouble in strong winds.

Well, interesting, not sure what to make of it all, maybe a past life connection for either my wife, Jamie or me, or all? I had a long talk with my wife about her affair with Wickham hundreds of years ago, but she is still denying it, well my investigation is not over, just on hold for awhile until I cross over.

One last major ESP event involving Zack and I occurred 7 months after he passed. The travel agent who arranged our Monroe training called me a few weeks after we returned and asked me for a favor. She had organized a weekend training in July 2013 for trainees wishing to develop their psychic skills, and she asked if I would serve as a training target for the 10 trainees. They each would attempt to discover things about me using ESP. As I had many interesting events in my life, I would be an ideal test subject. I agreed to do it, and each spent 15 minutes by themselves with me attempting to perceive facts about my life. The results were impressive, none of them knew me, or even my name, before they came that day. I was only to tell them if they were correct but not allow "cold reading" tactics of their asking vague questions first, then narrowing-in questions in response to answers, fake psychics do that, but these trainees had real skills as I discovered that day.

6 of 10 trainees mentioned my issues with my parents, a few those with my brother too. My relationship with him by then much worse after dad developed dementia about age 69. Dad slowly slid into, as First Lady

Nancy Reagan once said, "the long slow goodbye", ending in a long coma, dying at age 81 in the county nursing home nearby the Barn where I worked then.

My brother was no help over all those years in caring for our parents, and at the end, even insulted me by blaming me for dad's death, claiming I had not done enough to get him medical treatment for dementia. Him saying that to me on the night dad died causing me even more grief, absurd as there was no treatments for dementia back then, little if any even today. Of course that type of irrational behavior is due to his feeling guilt from doing little for his parents. Guilt produces anger at oneself, which is often projected outwards onto others. I understood that, but still I cannot forgive him yet.

One trainee said my mother is here now along with her two brothers and they are all proud of what the family has done to help others, that was nice to hear, and she did have two brothers. The trainee next said "your mom wants you to forgive your brother," wow, was that on target. But another trainee totally blew me away, she said your son is here now and he says you are thinking of writing a book about his visits, and he wants you to do it, and she mentioned I was thinking of writing an "e-book."

You see, just a few days before this training Jamie and I had a phone discussion about my writing this book, and I had even said maybe an e-book so I could include Zack's voice recordings as evidence, maybe they could be part of an electronic book on-line. This trainee had nailed it, a home-run, impossible to be a lucky guess. I soon wrote 3-4 pages, but sorry to say, no more writing again until seven years later and my life had now settled down.

Glad to report better news about my parents later in life, by their mid-fifties, their relationship with each other had thankfully improved. Dad lived at home again and thus my mother's mental and emotional health was better, as well, her issues were less noticed by others at that stage of life. When my kids were born, my parents provided endless baby siting for them and mom and her grand kids had wonderful times together, they love each other dearly, so a happier time for her, at least until dad's dementia.

A few years after dad passed, in April 2001, on a bright spring day, I had my mom over from her assisted living home. She wore the flowered blue dress I had given her for Xmas, I explained her savings were near gone and she had to move-out of the expensive assisted living where she had been for 7 years, but that I would move her into my house. She glowed at that news, evidence that I cared for her, and she was not to be abandoned. Then she surprisingly said, as if a premonition of her life's end, "it won't be long now," and fell asleep on my rocking chair on the deck.

I had the same thought as when last seeing Zack, I may never she her again. I stared as my mind caressed her image, as if my last, as I examined my complex feelings about her. My premonition, as with Zack, too soon fact, she died a week later.

Chapter 7

Finding the Spirit at the Barn

YSA had provided casework to school truants for decades, especially among delinquent youth who were required by court to attend school, but sometimes defied that. In 1991 a new YSA program was created as a mild sanction for these youth who had probation infractions including truancy. For many years the consequence to delinquents not attending school, or not complying with other court requirements, would be weekend lockup. But that was costly at about $900 per youth for weekend detention, as well, there were few beds available anymore. The Juvenile Probation Chief, Bill Ford, relied on YSA to help him solve problems, he was strongly committed to community-based services thru public-private partnerships, he inspired many, and still remains my mentor.

He called me back then and asked if we could take these delinquent kids into YSA's 3 local community group homes as weekend placements, but I had to say said no, the homes were near full already, and we had enough problems with our group homes in local communities. It's an impossible task to keep delinquent youth in group living homes from occasionally bothering neighbors, though mostly just minor issues, nearly impossible to eliminate, sometimes more serious issues too.

However, I offered another solution, YSA would take them backpacking for the weekend on the Appalachian Trail (AT) in the PA mountains several hours away. I had already been taking youth in my care on occasional weekend AT trips, at both YSA and in previous jobs. He liked that idea and it became a new year round program operating from Friday after school, to early Sunday evenings, called Weekend ACT. Eventually many other counties used this program too. Every weekend dozens of youth for 25 years straight, tens of thousands by my retirement, went camping and did community service at state parks, picking up trash or trail repairs on the AT. Winter trips were especially rigorous, Zack and Jamie went sometimes too.

Soon we needed more indoor sites during bad weather for this and other YSA programs, as well, more youth construction projects and community service to teach vocational skills. Some delinquents were paid for their work in our vocational programs, with the cash they earned used to pay court ordered restitution to their victims. This due to new **Restorative Justice** laws that changed court mandates across the county.

Over the years, YSA paid crime victims almost $1 million cash through YSA restitution programs. In later years, one restitution program was a bakery and deli we operated wherein 5 or more YSA youth learned skills while earning cash to repay victims, as much as $40,000 a year in wages earned by youth, plus state food services certifications like "Safe Serve." It was a self-supporting social service program rather than publically funded, and it never made any profit. It made food at cost for YSA, and provided food sales to the public. A novel, self-funding program generating both skills and funding for victim restitution.

Yet another need by the early 1990's was for indoor sites to increase our adventure challenge therapy (ACT) for all YSA programs. As stated before,

ACT challenges produce benefits that enhance more traditional talk therapies. By this time we had over 500 youth a day in all programs, and by 2001 we had 1,000 each day in all YSA programs. These were the good times, our high point.

We had built outdoor climbing walls and ropes courses in several locations, and the idea came to me that a large barn could be an ideal indoor ropes course and rock climbing site, allowing those activities no matter what the weather or time of year. And I knew where there was an abandoned, large historic barn, in a large complex of county buildings where YSA already had office space for many years, Neshaminy Manor Center.

Trust and support of YSA by court and county officials was strong then, but they may have thought me a little nuts as I spoke in 1993 about my vision for the barn. It helped that we had a good track record in construction, we had built a small house as youth vocational training, and then sold it and recovered all money spent. I offered to recycle that $125,000 of YSA income to start the Barn restoration, and so it got a verbal green light on trust alone. Abandoned since the 1950's and no longer a functioning farm barn, by this time it was just a place for the county to store seldom used and forgotten equipment. Doors hung open, wildlife lived inside, you could smell that, and the sense of history.

As we started the Barn restoration, some unexpected partners emerged who wanted us to use the Barn project as an alternative school. Local school leaders began discussions about the need for a new type of school due to the rash of horrendous school shootings across the country, such as the Columbine shootings. We were asked to create an alternative school for high-risk students.

It was clear when these shootings occurred that school staff already knew the shooters were high risk kids, an isolated, odd small group at school, but few on staff had taken the time to really get to know them or understand the real dangers they posed. What was needed was a new small school model that focused on student relationships with staff, and with flexibility to place students in it quickly, and later return them to their previous public schools without the legal process of having students withdraw totally from their home districts, up to then that was required for all transfers to any private school. This was a new kind of public-private partnership, allowing quick assessments of student needs and the risks. YSA's focus on "relational therapy" with student immersion in non-traditional new therapies put us in the forefront. During the multi-year Barn restoration, about 30 or more students from two local school districts, Central Bucks and Neshaminy, spent part of each day in classes and part working on the Barn. They helped as the high ropes course was strung high-up in 3rd and 4th stories of the Barn loft, and the 50' high Barn end walls became a simulated cliff for rock climbing. As well, a low ropes challenge course 8 feet high among the large oak beams supporting the ceiling.

We created classrooms in the basement were once the cows and pigs spent winter nights, though it took months to get the smell out, well to be honest, years. We added a music recording studio as part of a large computer lab, a science room with dissection tables acquired from a federal surplus program, and a multipurpose room with a large kitchen for school lunches.

The Neshaminy Manor Barn was built in 1732, at Rts. 611 and Almshouse Rd, near Doylestown, PA. Almshouse of course meaning poorhouse, and from 1806 into the 1930's the Barn was part of the Bucks County poorhouse and residents worked the Barn and surrounding farm for food. Decades later, the complex became a military hospital during the civil war, and finally a farm worked by inmates from the nearby county prison until the 1950's. We started the Barn project on faith, having no idea of the final cost, eventually reaching over $1.2 million after 5 years.

The Barn ghost had been seen by many others before I finally saw it too. Over YSA's 25 years at the Barn, some saw a tan colored mist moving up high and fast down a hallway, and that's exactly what I saw in 2015, though a few have claimed to see a formed body. There was nothing fearful, benign ghosts encounters were somewhat common in this large complex of historic old buildings. Even YSA staff and I had experienced them years before nearby, in Building G, the former civil war hospital, where YSA and Big Brothers & Sisters shared donated county space for many years in the 1980's.

In the 1830's, cholera swept thru Bucks County and of the 150+ residents at the poor house, over 80 died within weeks. Bodies piled-up as some poorhouse staff died too, and no one wanted to go near the bodies and risk infection. Finally a group of volunteers laid them all in a mass grave and most of them died too from Cholera, so no mystery why this area is haunted.

Once I had the security guard unlock the main door to the large, 4 story, G Building early on a Saturday morning, the entire building was locked tight and I was the first one in that day. As I reached the basement elevator to go up to my office on the top floor, before I even touched the elevator buttons, it descended on its own from above to meet me. I laughed and as I entered asking the spirit to take me up, but no response, I had to push the button myself, no humor. Another YSA staff came evenings by himself, and once he went to the same elevator but it wouldn't work, then he said a grey mist formed above him and a voice said "walk" … now that's funny!

By 2012, after the recession of 2008/9 our large alternative school at the Barn had closed, as had most of our other 13 schools, done in by eroding school budgets and competition from less effective, but cheaper, cyber schools that had be been expanding everywhere. Though I have read the average enrollment in cyber schools lasts just 3 month, high drop-out rates and even school violence were less of a priority after the recession. Funding cuts ended most all PA Alternative Education for Disruptive Youth (AEDY) schools in the state, we had pioneered the first AEDY school just 15 years earlier, now all gone, time to move on, we don't make public policy, and by then I had made enemies in both local political parties. As I said, I was good at that, as well rival companies always had their daggers out, hiring away our talent. Even in social services competition is brutal, but still a benefit, competition does improve quality and lack of it in the public sector is a weakness. Many stories I could tell, but not here, another time, in a book just about my experiences in social work, well maybe.

With an empty downstairs in the Barn again after the AEDY school closed, we decided to copy Jamie's successful Nature Center (NC) at our large residential ACT Camp in the mountains. It was already our most popular program ever, and very therapeutic. We would make the lower level of the Barn another nature center, and open to the public, and hope to attract children and families to an indoor educational mini-zoo. It could teach animal care and empathy to our delinquent youth who needed community service projects, they could help maintain the NC, and public fees could fund it. It was the win-win formula I always sought.

In 2013, as we coped with the loss of Zack and the onslaught of grief that followed, Jamie and I lost ourselves in creating this wonderful new program. My NC layout was to follow a path, a walk in the woods, leading from exhibit to exhibit, past waterfalls into fish and turtle ponds, and on to cages in caves. The concrete floor was painted into a stone path, I had mastered using spray foam to create large simulated rock walls and features. We built an aviary filled with small birds that would land on visitors and eat from their hands, we had macaws and parrots, one even entertains with music, everyone loves Elvis the dancing macaw, he even freezes, doing a "pose" to music, so cool. Plenty of reptiles, large and smaller lizards, various snakes, some rather large. Bunnies, ferrets, rodents and small mammals, and of course, another large tortoise, one now at each YSA NC. I think Darwin and Wickham would like the Barn NC, hope they visit soon, or I'll show them myself someday, Zack did say "seeing many people"!

Birthday parties, scouts and school trips kept the Barn NC busy thru the years, as well, popular week long summer camps with strong educational programs like stream studies were kids would examine the health and biodiversity of local streams. Upstairs the ropes course and climbing walls remained, now available to the public. The idea was a home run, in just a few years we topped 10,000 guests each year after. Seeing this public support renewed our family's spirit, and for me, faith again in my career, success was just what we needed.

Our family and YSA staff loved the Barn, but by 2016 we knew our management of it would soon be over. The Barn is owned by Bucks County, so never really ours, we were caretakers who restored it, with great benefit to our community. Though I retired from YSA in 2014, I remained part-time after that until 2017, helping Jamie and others struggle to keep the company going, after the great recession brought reduced public support, and with remaining funding more focused on maintaining the public sector services. Gradually many private programs withered or perished. YSA hung-on, but needed new leadership, Jamie overworked and me seeking actual full retirement.

By 2016 we had operating losses for years and our reserve funds were low. In late 2017 the YSA Board voted to hand the company, with all its remaining programs and still significant assets, over to a large conglomerate that was quickly acquiring by "affiliations" struggling smaller non-profits like YSA. We thought it was the best chance to save the remaining programs and the two NCs, but in the years following the merger, all but a few YSA programs were gradually closed by the new owners. Only the Barn Nature Center remains today and one other program, and now the NC's future is less certain too.

As I write this in spring 2020, the Barn NC's director just posted on Facebook notice that she was just told to close it quickly, breaking a merger promise to keep YSA programs open for 5 years, this just before, and not because of, the COVID 19 shutdowns. The public saw the Facebook notice and in just over a week there were 10's of thousands of views of that Facebook page and many calls to public officials to encourage them to save the Barn NC, with supportive comments, not surprising as the Barn NC for the last 6-7 years had so many visitors. It's loved by the community, and Bucks County leaders just responded and pledged to save the Barn NC, so we are hopeful. More history of Barn NC in Addendum, or see more of the NC at:

www.barnadventures.com

As we made arrangements in early 2017 to turn YSA over to the other management company, we wanted first try to document our Barn paranormal experiences. A professional paranormal investigation team was contacted and did a great job, with details also in Addendum. We now know for a fact that the Barn spirits appreciate its restoration, the programs there and the staff, they told us that directly during this investigation, amazing, Jamie and I were there to witness it.

Spirits were asked questions, whether they liked the staff at the Barn and the Nature Center. The spirits answered yes by lighting up a K-meter, or a flashlight, both common investigative devices. The K-meter measures electromagnetic energy near the device. It lights up in stages as stronger electro-magnetic energy is detected, the stronger, the more lights come on, moving energy sources believed to be from spirits, though a K-Meter can also detect high energy levels around wires and electrical devices such as circuit breakers, they produce steady lights on the meter. With ghosts, it is believed they let you know their presence by their coming closer to the device and causing it to light up.

We got strong responses to the questions about staff and NC, but more impressive was how the investigators even got the names of the ghosts that night. A group of 6-7 investigators, one an accomplished psychic who worked with police recovering the body of a boy who drowned in a local river a few years earlier. Jamie and I sat in a circle with them in the main Barn loft in the dark at night, investigators with the K-Meter a flashlight, a thermal camera, and a spirit box scanner.

The team leader asked "is anyone here," a light blinked yes, then he said "we want to get your name, does it start with A" the answer yes, "is the next name letter B", yes again, third letter C nothing, D nothing, he kept stating letters, we had Ab as the start of the first name. Then someone in the group yelled out "is your name Abigail", then all meter lights blinked on and off over-and-over many times, as if showing excitement that we had gotten her name.

Research later revealed that Abigail had been at the poor house in the 1880's, she had worked with the children, and in the investigation the team also identified a child spirit at the Barn too named Nate. An Afro-American child who loved to hang-out with the Barn animals, he and Abigail were often together, and still are now at the Barn with the animals in the Nature Center, so wonderful to discover. Had they earlier played a role in giving us the ideas for the Barn and the NC thru intuition?

But they are not the only spirits identified by staff that night, downstairs in the former animals pens, there was another, probably the one that I had once seen in that area. That spirit actually spoke through the spirit box which scans many audio frequencies continuously like a police scanner. It captured his German accent voice, as well, his apparition was seen full-body in the main hallway downstairs in the thermal camera. They asked him to come closer and after awhile he did, and they asked his name, the spirit box responded, Heinrich. County record checks later confirmed that Heinrich was the name of several nearby residents in the 1880's, maybe he worked at the Barn back then and he is still doing his job at the Barn NC today and that's exactly what he said to investigators. An amazing experience that night, with many orbs observed in the loft and shown in an Addendum photo.

Jamie seems to have visits from Zack, and I had a few too. She would see lights at her house flicker, and then asked to turn them on-and-off, and it would in respond to her requests repeatedly. Kia had a toy robot that would do the same, one time, I asked it 4-5 times to turn on and off and it did. However, ghost investigators warn that some spirits pretend to be others so we do not accept that as evidence of a specific person, only good validation of another paranormal event.

Once Jamie was driving and she sensed his presence, she asked out-loud "Zack is that you", no answer. Later she got a call from the psychic who had led the Barn investigation, they had became friends, the psychic said, without Jamie telling her of her premonition in the car, that Zack had been in the car with her recently and she had known that. Wow, we are blessed to have these confirmations occur over and over.

One last spirit event was special for me, everyone knows I love dogs and have had one most of my life. My wife complains I hug the dog more than her, probably true. My White Lab, Oli, was a 90 lbs. poorly trained beast of a dog, whom I loved dearly. He would snatch a whole pizza off the table if left alone, gobbling it all, once even a whole chocolate cake, another time he swallowed a whole baby rabbit.

Oli had been put to sleep a few months before we lost Zack. He was age 15 and could barely move by then, incontinent for over a year, his time had come, and my nephew did me the favor of taking him to the vet. that was too hard for me. By then I had become the caretaker of Zack's puppy Kody, a Golden Retriever still with me today, watching me now.

A few months after losing Zack, he returned for a brief visit to let me know Oli still exists too with him. In the spring of 2013, while driving back to my retirement home FL from PA, with Kody as company, I was in my camper and spent the night at a rest stop on I-95 about half way. In the morning, just at dawn when I was semi-awake, I heard a dog bark inside the RV, not Kody who was next to me, he stirred at the sound too, it was farther away, it was Oli's distinctive low bark, and it was followed quickly by Zack's voice saying just "Dad".

Zack had brought Oli to me, I am so grateful, and Oli was to bark again, just once at dawn, a dozen or more times over years. Though this was the only time I heard Zack speak. So dogs become spirits too, comforting, but I wonder when I pass, will all my past dogs greet me altogether at once when I cross-over? Unlike other events, I have no evidence to show, but that was special gift from Zack to me, thank you son!

The Almshouse Barn Restored by YSA

Chapter 8

Can Science Explain the Paranormal?

From Quantum Mechanics to
Holographic Universe & I Love Lucy

I have some background in science and have tried to follow the emerging insights of physics over my many years since college, but still have only modest knowledge, and my distain for math doesn't help. That said, I see areas where current theories of reality can offer insights into how paranormal events exist within the laws of nature. The CIA's declassified research on Monroe Institute is a source of information here, as well many articles and videos I've studied. A summary from these sources is that matter, and therefore our perceived reality, is an illusion. Actually, objects we see arise from ripples or vibrations in the underlying energy fields that are the real reality. That's also foundation too of an emerging recent theory- the Holographic Universe.

Many physicists today accept the above, that reality is at the most fundamental level, patterns of vibrations in energy fields. Then it follows that paranormal events too are patterns within those same energy fields, with the addition that some energies are conscious. Spirits and humans have a basic similarity, both exist as part of energy fields, and they can be self-aware.

Our brains and thoughts are much more than the bio-chemical reactions of seemingly material brain cells called neurons. What travels thru our brain cells as thoughts, perceptions and memories, are actually patterns in energy waves. As well, our mental energy waves also expand outside our bodies, they interact and are part of external energy fields. Recent research shows that energy waves communicate throughout our brains without any physical connections within brain cells, again non-physical communication of information carried as patterns in these waves. When information is exchanged between entities through these wave patterns, that is called telepathy. That can be either human to human, spirit to spirit, or a mix of both.

Biologist and researcher Rupert Sheldrake believes that conscious energy fields surrounds and interacts with us and with animals too. His research shows when a flock of birds all turn at once, that is by telepathic communication. Not the old explanation that each bird watches the birds next to it for movement cues, that would take longer than observed in flock behavior according to Sheldrake.

Fields of consciousness energy is an idea proposed by psychologist Karl Jung many years ago with his theory of a collective unconscious. Today with field theory accepted as the fundamental reality, the link between everyday reality and paranormal experiences does not require an abandonment of science, but fits well within it. Why some of us experience more paranormal may be that our minds are better able to "tune-in" to the frequencies of vibrations in energy fields around us. In other words, more sensitive receivers, but everyone has this same potential capacity. For more, see U-tube videos explaining reality by physicist

Tom Campbell, formerly of Monroe Institute, as well biologist/researcher Rupert Sheldrake.

Are there things we can do to improve sensitivity and awareness of the paranormal? Yes, that is the training provided by Monroe Institute called "Gateway" and studied by the CIA, you can read it online, declassified in

Sept 2003: **CIA RDP96-00788R00210016-5**.

Gateway, as discussed earlier, is a training we attended that is available today as an online course. Mediation is taught with the use of Hemi-Snyc auditory technology to entrain and synchronize brainwave to reach the deepest levels of consciousness. In the deepest states, communication with conscious energy fields around us is enhanced, as too are paranormal experiences.

After our Gateway training, I mounted a stereo CD player on the wall above our bed with the speakers on each side of the headboard. We played the Hemi-Snych stereo CD every night, almost continuously, for 4 years. They sell other meditation CDs too, one called "The Journey" is especially powerful.

Hemi-Snych has restful background sounds along with the Hemi-Snych beats which are different for each ear. However my habit of sleeping on my one side and flipping from side-to-side all night probably hindered my maximum benefit. We experienced this brain wave entrainment for a long time and during that time our paranormal experiences continued and increased, so I accept that it was helpful.

We had more lucid dreams during those years, in one I was with Zack and hugged him while crying. I could feel the contact as if a real experience even as I became fully awake. In another dream I was talking on a phone, a garbled voice on the other end, then it came into focus and was Zack's voice saying again "seeing many people". I sat up quickly, still hearing him while awake for a few seconds more. Were these dreams just wishful thinking, dreams, according to Psychoanalytic Theory, are rooted in our desires and fears. I missed him, so creating wish-fulfillment by dreams is a natural expectation. Whether experienced by dream or actual spirit visitation, I am grateful to have had them.

Any discussion of the nature of reality for well over a hundred years must include the famous double slit experiment. It exposed a great mystery of reality, a paradox, not understood to this day. You already have heard of this mystery, its called Quantum Mechanics, the physics of the very small, as opposed to Einstein's General Relativity, the physics of the very large.

Do not feel bad that you can't understand it, no one does, there are theories, but no agreement yet on which are correct, if any. It deals with the basic question we just discussed, is reality waves of energy or actual solid particles (matter), and it confirms that it is both at the same time. When your mind perceives it, it appears as particles, seemly discrete small objects, but when you don't look or measure it, reality is waves. This experiment detects these states. The double Slit Experiment for beginners like me, and probably you, is best explained by animation, so a Google Double Slit Experiment Cartoon, or see it here:

https://video.search.yahoo.com/yhs/search;_ylt=AwrCmmbZNn-BeFUoAXwAPxQt.;_ylu=X3oDMTByMjB0aG5zBGNvbG8DY-mYxBHBvcwMxBHZ0aWQD

For those who don't like cartoons, here is my simplistic description. At one end of a room you have a device that can shoot single particles, such as single bits of electricity (electrons), or bits of light (photons). The device shoots them across the room at a screen or target that can show where the particle hits it. Then in front of that screen place a small wall that blocks the particles except at two open slits that allow some particles to get thru either slit. Behind the slits is still the screen that shows which slit they went thru, unless they we fully blocked. Do this over and over and you see a pattern on the screen of where the particles land behind each slit.

Now forget shooting particles, and consider sending waves instead, waves as in the ocean sent through the same device. Though using water is not part of the actual experiment, only for my description here, visualize the room with the same devices half submerged, then send water waves across the room and thru the double slits with them registering where they hit the screen behind the slits. Waves travel across the whole room, so will always go thru both slits at the same time, unlike single particles that go thru just one slit or none. So waves show a much different pattern on the target screen than particles, that's how you can tell which type went through.

That's because a single wave goes thru both double slits, it leaves from both as two waves behind the slits, one from each. The two waves then run into each other and interfere with both size and shape of the waves. If a high and a low wave sections hit each other they cancel each other out and there is no wave, but if they merge at peaks the resulting wave gets even higher. Just like waves in the ocean do when two boats have their waves meet and interfere with each other, if you ever see that, watch the wave patterns closely. The peaks and valleys inherent in waves create an "interference pattern" on the target screen in the experiment. So remember that particles and waves leave much different screen patterns in this experimental device.

Now the double slit experiment paradox, if you look and measure which slit a particle of electricity or light went thru, the target screen pattern will show it went thru only one or the other of the two slits (or neither) in other words, a particle pattern. But if you don't look and don't measure, it goes thru both at the same time just leaving a wave pattern on the screen. Matter is both particle and wave, depending only if you looked or not- confused ?

What we learn is proof that reality is different when we observe it from when we don't, and consciousness is a key part of the process needed to experience everyday reality. Energy waves create the perceived reality in our minds of matter, and reality is matter only when minds perceive it. Yes, we are all confused!

Science has done well to define the laws of particle physics, that's the science of every day reality. Wave science is less understood and is also the science of both reality and the paranormal. Waves are depicted on paper and in electronic oscillators as a pattern of squiggly moving lines with peaks and valleys. The distance from highest top of the wave to the lowest bottom point is called the wave amplitude (the "A" in AM radio). The distance between moving waves is their frequency (the "F" in FM radio). So amplitude variations (wave height) and frequency variations (spacing of waves) are key characteristics of all waves, and therefore of the reality of energy fields that create our universe.

The CIA report proposes something very strange. At the tops and bottoms of energy waves, before waves turn from one direction to the other, that the ends of the waves leave our reality. Our reality maybe has amplitude limits in its waves, just as our hearing of sound waves have limits to frequencies. So as the fabric of reality vibrates, maybe waves leave our reality briefly, and interact in unknown ways with another reality or even multiple ones, well, not sure that's been proven.

But it might explain why paranormal events seem not to exist for long, but come and go quickly in our reality. When Zack said in his recording "I'll visit you at the house" he is not talking about coming from somewhere nearby in our world, but from another place where they exist after death. They cross back and forth to visit. As well, when Zack says "seeing many people" and "its so perfect" he is describing his new spirit reality just after he died.

An example to help you understand and accept this scientific conclusion that perceived reality is not as it appears. Let me expose just one illusion everyone experiences all the time, and that you are experiencing right now reading this.

As you read this you are probably sitting on something, the paradox is that you are not really touching anything. You are not actually touching the book you hold (or the computer), or even whatever you "feel" that supports your body, that's all illusion. Matter never touches other matter, never as the underlying energy fields repel each other. No real contact with anything in your whole life. The fields get close, interact but never touch, but you think in your mind that they do.

Lets look closer at this, touch something now with your hand, say a table, what really happens at the atomic level? As the atoms and their energy fields within your hand get closer to the atoms and energy fields of the table, the energy in each repels each other. The harder you push the harder the energy fields push back, you can't force these fields together and make contact. Contact is an everyday illusion.

Lets discuss reality as energy waves and not matter. An example of energy waves are the broadcasts of TV shows you watch (for sake of this argument ignore cable TV). Now lets assume you are watching a re-run of your favorite "I Love Lucy" show (or for some your favorite cartoon show), would it make any difference if you place the TV set up high near the ceiling, or on the floor by the window, no. The point is the TV waves are everywhere in that room, and outside, not just in one fixed place, that's called non-locality. These moving TV wave signals are converted by the TV set into a moving visual image of a particle-based illusion of the everyday world.

TV shows are fake reality, illusions perceived by your mind, but created from information embedded in energy waves traveling from outside the TV set. I think that is a helpful comparison. Our minds are like TV sets, they receive information thru energy waves from your senses, then construct in your mind images of objects. Sounds like the Matrix Movie, doesn't it, though no one thinks you have a body somewhere linked by brain implant to a computer, well, almost no one thinks that.

Micro-second to micro-second our reality can vary, it may appear different as energy fields are not static or fixed, but move and change. Monroe Institute teaches that we all have psychic abilities and more than just five senses. I believe minds exist within these universal energy fields, and within them we have access to other consciousness around us. We are like characters in a video game, and when we die in this illusion we live in called reality, it has no effect on our mind's existence, it still continues on as part of energy fields around us. That explains our telepathic communications with spirits, and with Zack's cell phone messages to Jamie, his energy waves directly recorded by the cell phone, just as electronic voice phenomena (EVPs) and spirit voices are recorded by many investigators with digital or magnetic tape recorders.

Intension and intuition are key aspects of our minds, and prayer expresses our intensions to others, human or spirit, thru telepathy. We are all always linked to these fields surrounding us, Karl Yung's collective consciousness, and Sheldrake's theory of Morphic Fields are correct. We get information and knowledge from these energy interactions, **and that is where intuition comes from.**

The theory of the "The Holographic Universe" is summarized by Michael Talbot in his book **The Holographic Universe**. It explains the illusion of material objects are created by energy waves called holograms formed by two intersecting waves of light. Intersecting energy waves may also be how our minds store memories and manage sensory perception and thinking.

Among Steven Hawkin's final research included that the universe, thru time, becomes just information stored in 2 physical dimensions at the event horizon of black holes. He thought black holes are where matter leaves our reality. In a way black holes are like hard disk storage devices in a computer, they store vast amounts of information and that information is a wave energy field.

A hologram is a beam of laser light split into two beams with one beam then bounced off an object. When the beams are later crossed again, where they intersect, the object appears there as if real. For those who have visited Disney's Haunted Mansion ride, the three dimensional ghosts there are created by two intersecting light beams-they are holograms.

Holograms can store vast amounts of information, just as our minds do. A recent theory is the energy fields underlying all reality are holograms, generating all we see, if true, everything in the universe is an illusion of light. The vast information capacity of holograms may explain how the universe seems so vast, though not the why.

In summary, we live in an illusion, we are not really matter, nothing may be, objects are illusions, everything is actually vibrating energy fields, perceived in our minds as objects. In death the mask is removed and we join with other energy consciousness. My most valued conclusion from all the paranormal events just presented, as well the science just summarized, is that there is no death for any of us, just transformation back to conscious energy. It all supports the evidence that Zack and Sean are still with our family, they exist, and they do visit us, as affirmed by the many ESP events just reviewed, voice recordings and photo evidence. Your deceased loved ones exist too, and may visit, just ask them, they may hear you.

Our family prayers flowed from our grief and demonstrated our strong need for spiritual reunion with our loved ones. We needed that to continue our efforts to help others. Though we lost them, **our family were given a very special gift, and we are glad to have shared it, and hope now to pass this gift of awareness on to you.**

Zack 2010 at a South American Animal Sanctuary with Jamie

Addendum A

Youth Services Agency & Community-Based Services

The Supreme Court's famous "Gault decision" in 1966 had forced changes in juvenile justice everywhere. It was a case with the underlying "holding" that making kids a delinquent often further harmed rather than "reformed" them, as the law required. Gault was a 15 year old who was only accused of making a prank phone call in AZ, with no evidence or due process protections given all adults, he was committed to a state reform school until age 21, a six year sentence. The Supreme Court reversed that and found that the juvenile justice system needed to change, and to provide youth the same due process given to adults. Other court decisions also challenged juvenile courts, finding that services intended as rehabilitation were as often the opposite. Behaviors like truancy and running away should not set in-motion legal processes that lead kids into harmful confinement.

Status Offenses

These court rulings meant that a class of youth behaviors that once were crimes and leading to delinquency convictions and lockup, were no longer criminal. The name "status offenses" meant they only applied to minors, these are: truancy, running away from, and being ungovernable by their guardians. Across the country by the early 1970's new youth service bureaus were established in every jurisdiction, and in Bucks County, PA, their new program was Youth Services Agency (YSA), my new employer in 1976.

YSA provided crisis intervention and tried to stabilize problems like running away, truancy, and conflicts and fights with youth and parents, then a referral to longer term family counseling or other type of help. Therapy helped families define and accept family rules and use words rather than deeds to solve problems. But youth who remained uncontrollable or truant were a challenge. After several years as a casework supervisor at YSA, I became the CEO as I had a good reputation among police, school and court officials. They knew if they referred a family, we would respond quickly, stick with it, and try hard to help.

Crisis Intervention & Casework Techniques

We established protocols through trial and error, and realized that eventually conflicts between youth and parents disappeared once a youth approached age 18, when they were legally free to leave home and parental support can end. I helped them realize what life would be like with no parental support, something youth seldom contemplate. Truancy among 17 year olds was much different than among 14 year olds. Crisis intervention means quickly meeting with parents and youth all together, then separately, to understand issues. During the one-on-one with youth, I asked them their version, and also their plans for the future, which was often stated as a desire for independence.

I explained my job was to help them reach their goals at age 18, but until then, they should not destroy what positives they may have had over their lifetime. Conflicts now near the end of parental control were temporary, they would gain freedom in time (for better or worse). Right now it was in their interest to make peace, as we discussed what they needed to reach their goals. In short, I used advocacy to gain their trust, and I helped parents accept that gradually letting go of control was inevitable.

This approach was often an effective band-aide, putting a crisis on pause, then looking to family therapy elsewhere for more in-depth work. Using advocacy for youth, I gained leverage to get some cooperation, as well both parents and youth assumed I had backup from police, that mostly was a mirage, but it helped me, I used that, and police and court did support us.

Substance Abuse

Some family conflicts involved youth drug or alcohol abuse. YSA was not a drug intervention program, but police and schools turned to us when they got calls from worried parents who learned their kids used drugs, some serious, some less so. We sent all who would go to drug treatment programs, but to remain in treatment meant the youth had to agree to stop using, and to follow other 12-step mandates. Too many youth refused those rules, and were shunned as a result by some drug treatment programs. There was no solution to impact those kids who kept using drugs, little else but YSA to keep trying. In severe cases I would setup drug busts to force high-risk youth into treatment, but I couldn't do that for too many, it was not our YSA mission to turn kids into delinquents.

The public administrator overseeing the county drug treatment system became an adversary, as I went public with the fact some highest risk youth were not "treatable," it undermined the public belief in treatment effectiveness. Drug treatment had been a political fight for years to force medical insurance providers to pay for it, and as treatment programs became very highly regulated, they presented they had the "cure for this cancer."

Publicizing poor program outcomes is an existential threat, creating enemies quickly, especially in the public sector, and with my limited respect for unearned authority, I made more than a few enemies. I published research in 1984 and again in 1992 to explore YSA's experiences with treatment outcomes. With the country facing an ever-increasing plague of substance abuse, and with the treatment not as effective as hoped, a new self-help approach grew among parents and began to fill that void in the 1980s. **Tough Love** was invented in our region and became a national movement and public icon. A parental response to kids that wouldn't cooperate with drug treatment, it filled that gap.

Community Service and Restitution

Our programs had been supported locally by some Quaker leaders, allowing YSA to use an old, vacant, Quaker school for free as a girls shelter. I respected their core belief that God was for each individual to define and is within everyone, therefore no one's beliefs should be elevated above others. Their focus on community service was something YSA copied on a large scale as we became a leader in the new "**Restorative Justice**" laws adopted by many states in the 1990s. Replacing the older **Rehabilitative Justice** mandate that the court was to cure offenders. Instead Restorative Justice focused public resources to help restore the victims and the community. Youthful offenders were to receive skills, as opposed to trying to "cure" them of a disease. YSA provided means for youth to pay cash restitutions to victim and juvenile courts. About $1 million was paid as well ½ million hours of community service provided over the years.

In the years following the 2007/8 recession, states and courts cut funding to many youth programs, especially residential programs that cost as much as a college education for each youth in placement. The most expensive were behavior health residential "treatments" that were still trying to "cure" youth of complex social problems like delinquency, mostly by using talk-based therapy with very weak outcomes, they were using the wrong model in my opinion. However, the federal government would pay for it under Medicaid, as many kids in public care would automatically receive federally funded healthcare. So states spent freely on behavioral health treatments with little direct cost to them up to the recession.

Behavioral health cost 2-3 times that of YSA residential programs, with no greater benefit. As you have read, we explored new ideas in our search for effective outcomes, from Adventure Challenges, Opponent Process and animal care therapy, creating self-esteem, as well teaching positive values thru community service, and all at a lower cost.

Addendum B

Opponent-Processes Therapy

YSA developed our adventure therapy using stressful activities that create positive emotions, a non-verbal type of therapy. If clients did the activity they got the benefits, impacting resistant clients who may not benefit much from talk therapy. We went backpacking, used indoor and outdoor ropes courses and rock climbing, offered white-water kayaking and rafting, and even trips on our 78' three mast sailing ship, Lioness, with real cannons on the Chesapeake Bay.

I had learned in graduate school that opponent processes create the pendulum effect found in many human emotions due to homeostasis, the brain attempting to balance the chemicals that regulate emotions. If you push your emotional pendulum one way, for example, by falling in love, or by having fear thru rock climbing, when the experience ends, your emotional pendulum does not go back to neutral as before the experience. They go instead to the opposite emotional state, love brings heartache, and fear brings euphoria. You have no choice, it's wired into your brain.

Chemical transmitters such as the stimulant dopamine, and pleasure transmitter endorphin, an opiate type drug, keep in balance with GABA that dampens emotions, as well the chemicals behind fear being balanced by the opposite, those behind euphoria. These chemical opponent processes are also what happens when you use psychoactive drugs, the high leads to the opposite state, unpleasant withdraw. I published my research on the use of opponent processes as therapy as a treatment for resistant youth in the Journal of Juvenile & Family Court, Nov 1992.

Nature Centers - Animal Therapy

YSA's large residential Act Camps were off-shoots of the weekend ACT backpacking program, and was used for youth unable to remain at home, they lived at the Camp for a few months, sometimes longer. In my youth at summer camp, I was drawn to the Nature Center at Treasure Island Boy Scout Camp in the middle of the Delaware River. From those fond memories, and with the energy and ideas of Jamie, who had traveled the world volunteering at animal preserves, we created the first YSA Nature Center (NC) as part of the ACT Camp and near the AT hiking trail. It was an instant success, with a bird aviary, snakes, rodents, small mammals, reptiles, and a barn yard with pigs, goats, peacocks, turkeys, chickens, etc., all living in harmony together, (most of the time). I loved watching the giant pig roam the yard with 3 or more turkeys riding on its back, seemly enjoying the ride.

The Camp kids maintained the NC as their community service, gaining animal care skills. Those who had more interest could receive a 120 hour state approved certification to be an Veterinary Assistant, often leading to real jobs in a Vets office, the SPCA, Pet Smart, and other pet related businesses. Over a hundred YSA kids got that training. We soon learned some even wanted to come back to see the animals again after they left and went home. Often years later, they called Jamie, asked to stop by with their friends. That is unheard of for programs that youth are forced into by court. YSA ACT Camp was a staff secure residential program and they wanted to return again as they missed it, that is an unheard of compliment, the best kind.

Even better, Jamie published research on NC outcomes that document client gains in empathy produced by the program. Lack of empathy is a major cause of anti-social and criminal behavior, and there are no known treatments. Today we see the TV show "Pit Bills & Paroles" also demonstrate that caring for animals is a means to generate love and bonding with others. Jamie hears from past clients all the time, and her published research had among the highest viewership of any article in the Journal of Juvenile & Family Court in 2006.

Addendum C

History of the Almshouse Barn

Our first major building project was the restoration of the abandoned, historic, county owned barn built in 1732, near Doylestown, PA. Formerly a private farm with outbuildings and a manor house, called the Rodman Estate, then purchased in 1806 to become a the County Alms-house (poor house), with a large 4 story stone building built soon after the purchase to hold 150 residents. In the 1840's Cholera came to this region and over half of the 150 people in the county poor house died within weeks, many buried in a mass grave onsite. It operated from 1810 to 1966, then the poor house was closed and demolished and a county nursing home was built in its place. By 1966, welfare payments became widely available to assist the homeless live more normally in communities, as well as expansions of public housing projects.

The barn and surrounding farm was worked by the Almshouse residents to produce their food. The large barn was over a 100' long and 60' tall at its peaks. By the Civil War, the county farm also supported a new military hospital, called today building G. In the 1900s, the Almshouse farm also became a prison farm, worked by inmates of the nearby county prison. In disrepair after 230 years of service, the Barn was abandoned sometime in the 1950s. YSA moved nearby in Building G in 1984 and started the Barn restoration in 1993.

The paranormal investigators in 2017 identified by name 3 spirits currently at the Barn. Heinrich was found in the basement where the large NC is located, he is assumed to have been an animal caretaker of the cows and pigs in that space.

Spirits named Nat and Abilgail were identified by information from yes-no questions answered by spirits using the K-Meter and a flashlight, as well the psychic leading the group adding more information. In the Barn photos the number of large and bright orbs caught by camera may provide more documentation of spirits, there were over a dozen shown, some very bright and colorful, not dust or bugs.

I am proud of the many achievements of all who worked at YSA over the years, and of our new ideas and unique programs to help youth. I am especially thankful for the support we were given, by those alive and maybe by spirits too. The Barn remains, as it always has, a major support for community life, both human and animal, in the fullest sense. At my retirement party at the Barn, I told the guests about the ghosts we found there, and I promised them someday my spirit too would join these others at the Barn, and be among the spirit caretakers watching over this special place.

Kia (facing) with her cousin, Teagan, at Barn NC 2016

Descriptions of Photos - In Order

Pinnacle Overlook - Near Hamburg, PA a side trail off the main AT

Spirit of Boy - Semi-transparent behind the girl, its casts a shadow and partially blocks wall grease board. Note the spirit has a flesh color face, with boy facing towards the wall, you can see flesh color face, eyes and nose, etc. Close-up photo too

Barn Adventure & Nature Center - Rts. 611 & Almshouse Rds., Doylestown, PA. Photos show the ropes courses and some of the climbing walls and front view.

Zack with Owl - Zack around 2009 with Jamie volunteering at animal rescue sanctuary in South America

Kia 2016 -.Kia with cousin, Teagan at Barn NC, Kia seems jealous of the snake paying attention to her cousin.

Marly at Awards Ceremony - Two photos taken seconds apart, one with hands raised, the next arms become transparent (they are still raised though transparent. The 3rd photo a close-up of what's in front of Marly. No explanation, there is something in front of her.

Debra in yoga pose - with green, complex orb perfectly centered above her

Colored and bright White Orbs in the Barn Loft - taken by paranormal investigation July 2019

Barn Nature Center - Photo of the path, a simulated "walk in the woods" thru exhibits, I used spray foam to create appearance of rocks near animal displays.

Jamie - Photo with her big colorful MaCaw.

Marly & Prescott in Barn NC Aviary - About 2013 at Barn NC in one of the bird exhibits.

Lioness, YSA's Tall Ship - used mostly in the Chesapeake Bay for weekend trips, especially as a reward to those doing well in our ACT Girls Camp.

Colored Orbs at Barn Paranormal Investigation 2017

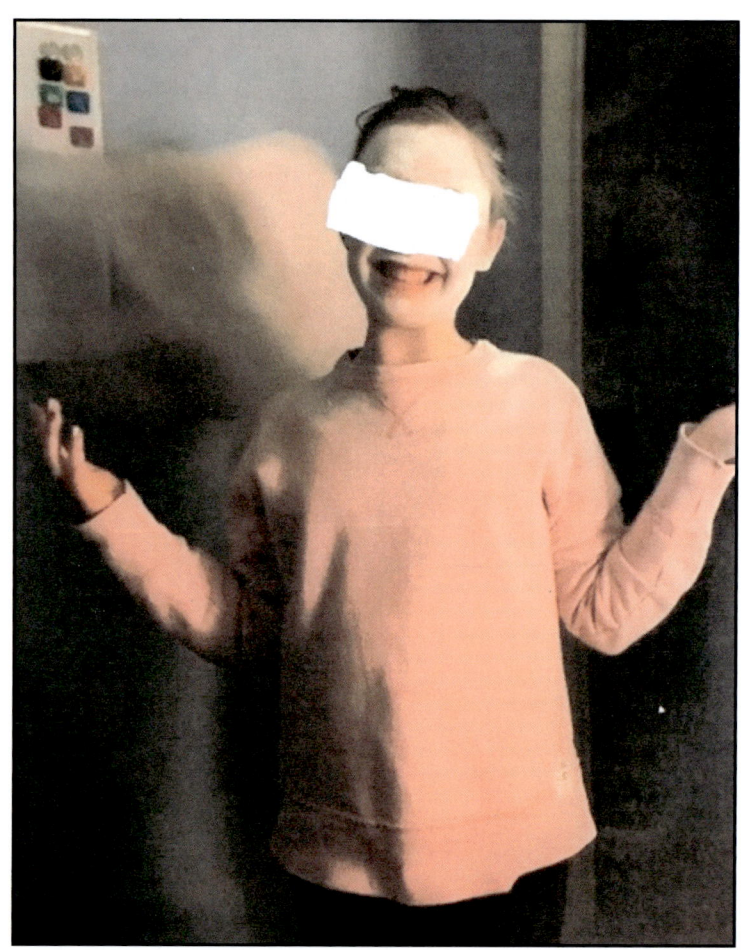

Spirit Boy Behind Girl

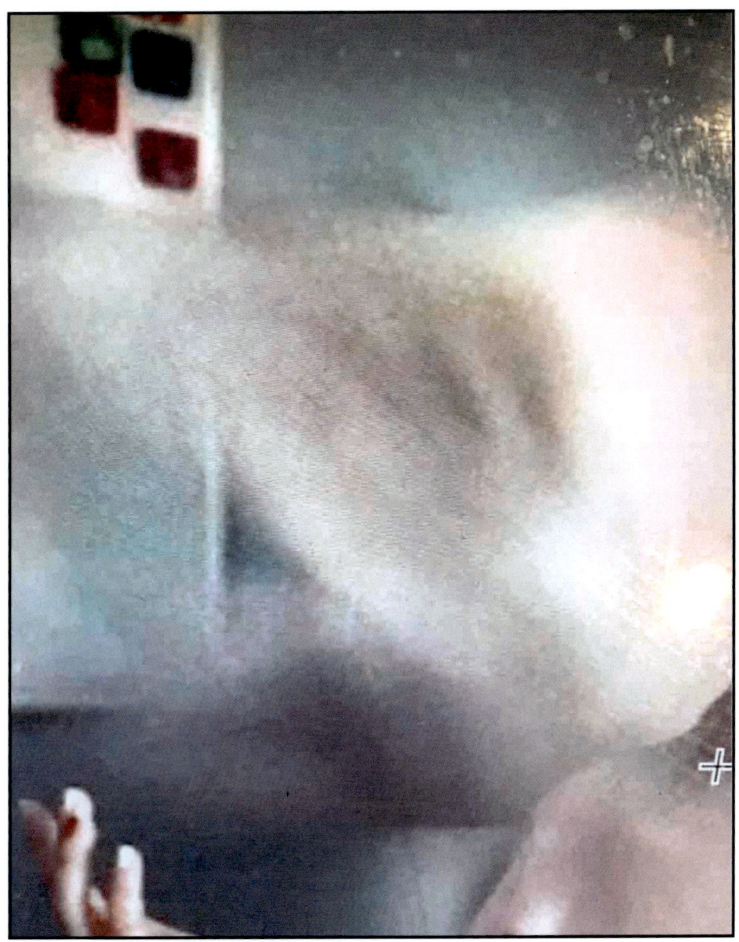

Wings? Notice Boy's Skin Colored Face

Marly at Gymnastic Awards 1st Photo (below)

But to left, 1 second later with arms invisible, something is in front of Her.

Debra with Orb

Barn Nature Center, Path Thru Displays, Fish Pond in back

Barn Climbing Wall & Ropes Course

Prescott & Marly in NC Aviary 2013

YSA Tall Ship Lioness

Jamie at NC